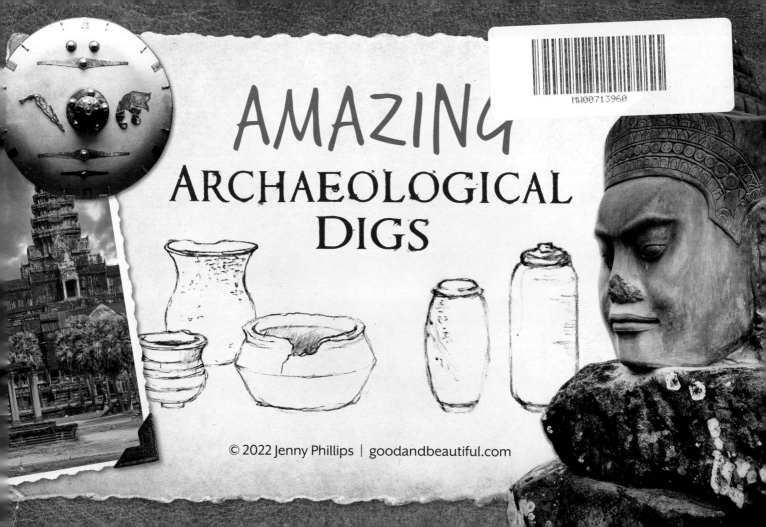

AMAZING
ARCHAEOLOGICAL
DIGS

AMAZING
ARCHAEOLOGICAL
DIGS

Written by The Good and the Beautiful Team

Designed and illustrated by Robin Fight

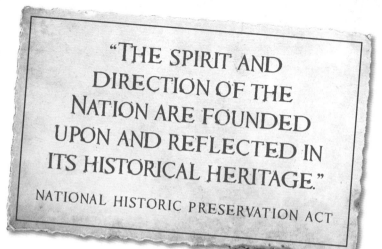

"THE SPIRIT AND DIRECTION OF THE NATION ARE FOUNDED UPON AND REFLECTED IN ITS HISTORICAL HERITAGE."

NATIONAL HISTORIC PRESERVATION ACT

ENACTED IN 1966 ENACTED IN

In 1966, the United States government passed the National Historic Preservation Act, which protects sites that are deemed important to the history of our country. One of the largest countries in the world had declared a stance on the significance of discovering and studying how groups of people who came before us lived and the significance of protecting these historical sites for future generations to study.

The hard work and determination of countless adventurous archaeologists to uncover and examine evidence of the lives of ancient and more recent civilizations have led to a deeper understanding of humanity—where we came from and where we can go from here. Knowing how people once lived, how they used the natural materials around them, what they did for jobs, what they ate, what sicknesses affected them, and what they treasured is essential to the understanding of how and why human behavior has changed over time.

Around the world there are many fascinating, mysterious, and sometimes slightly disturbing archaeological sites. Let's explore some of the most famous amazing archaeological digs!

ARCHAEOLOGICAL DIGS SITE MAP

NORTH AMERICA

Mesa Verde

EUROPE

Must Farm • Lindisfarne
Sutton Hoo • Rooswijk
Lascaux Cave

Hissarlik
Jerusalem Tunnels • Beit She'an
Thonis-Heracleion, Canopus • Dead Sea Scrolls
Pyramids of Giza • Petra, Jordan
Abu
Simbel • Tutankhamen's
Temple Tomb

ASIA

Terracotta Warriors

Ajanta
Caves
• Angkor

SOUTH AMERICA

• Machu Picchu

AFRICA

AUSTRALIA

Easter Island

THE PYRAMIDS OF GIZA

Few buildings are as recognizable as an Egyptian pyramid. While many pyramids have been discovered and explored in Egypt, the grandest of all are the Pyramids of Giza. These three massive structures have survived for thousands of years and are still items of fascination for many. The largest of these pyramids, the Great Pyramid, is the only Wonder of the Ancient World that still stands today.

These pyramids were not discovered in more modern times as many famous archaeological sites were. Instead, they have been known since their creation, but people's interest in them hasn't lessened. The Pyramids of Giza were built for Egyptian kings Khufu, Khafre, and Menkaure. The base of the largest pyramid is around 230 meters (756 feet) long on each side and used to be nearly 150 meters (492 feet) tall. The smallest pyramid's base measures 109 meters (356.5 feet) in length on each side and is 66 meters (218 feet) tall.

The Great Sphinx and the pyramids

Giza

EGYPT

Khufu

Khafre

Menkaure

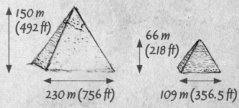

150 m (492 ft)

230 m (756 ft)

66 m (218 ft)

109 m (356.5 ft)

Why are these pyramids shorter now than they once were? Each of the pyramids was originally covered in white limestone. These outer casings have since been removed almost entirely. Robbers plundered the pyramids from the outside and inside in centuries past.

Though the Pyramids of Giza aren't in their original splendid condition, they do have educational value. Art in the tombs shows ordinary people living their lives—farming, taking care of animals, and participating in religious rituals. Artifacts also teach about ancient Egyptian life. Much can be learned about the Egyptian language, too, when studying the texts found in the pyramids. These pyramids and the areas surrounding them have been explored and studied by many scientists and other people since at least 1798.

Couple harvesting papyrus

PETRA, JORDAN

The Bible tells us about Moses passing through a valley and striking water from a rock while leading his people to the Promised Land. Hundreds of years later, groups of nomads began to settle in this same area, creating a powerful, independent kingdom through which international trade caravans carried gold, fine silks, and exotic spices. And then, in AD 106, Rome took possession of Petra, trade routes changed, and the ancient city was forgotten.

It was not until 1812 that the mysterious city was rediscovered by Swiss geographer Johann Ludwig Burckhardt. News of his discovery began to spread, and more people began to seek out the ancient ruins.

Al Khazneh (The Treasury)

JORDAN

• Petra

Petra is one of the most well-researched archaeological sites in the world. Its rosy sandstone rock facades, tombs, and temples landed the once-secret city on the list of the Seven New Wonders of the World.

Structured research began in 1896. Many important sites were identified, towering rock facades were classified, and over 800 rock-cut items and inscriptions were documented. In 1929, George Horsfield and Agnes Conway began archaeological excavation work, uncovering many tombs and houses. They were joined by a famous American archaeologist, William Foxwell Albright, in 1934, and together they excavated a site in the northern part of Petra, known today as Conway Tower.

After WWII, additional excavation work began, uncovering the main street and many shops. The "Arched Gate" of the temple, parts of the city wall, and valuable pottery also were found. In the early 1960s, the main theater of Petra was excavated, followed by the "Temple of the Winged Lions."

Main Theater

Temple of the Winged Lions

Jordan postage stamp, 1954

In the 1990s, the Byzantine Church was found with well-preserved floor mosaics. Excavations continue to be carried out by many researchers and archaeologists, many of whom have uncovered more history and new artifacts to be investigated.

Byzantine Monastery

Interior carving of Ramses II

Inside the temple are engravings of Ramses II and his favorite wife, Nefertari, along with drawings depicting Ramses' victory at the Battle of Kadesh. The artwork, history, and awe-inspiring large-scale size of this temple make it the second most visited site in Egypt, behind the Pyramids of Giza.

ABU SIMBEL TEMPLE

EGYPT

Abu Simbel

Johann Ludwig Burckhardt struck archaeological gold again in 1813 when he discovered the top of a deserted temple in the deserts of Egypt. Towering over the west bank of the Nile River, carved as one solid structure into the sandstone, are four likenesses of the Egyptian king Ramses II, two on either side of the entrance to the Abu Simbel Temple. Ramses himself commissioned the grand statues to be built, complete with smaller figures of his wife and children carved into the stone at his feet.

The temple fell into disuse and eventually was almost completely covered in sand, not to be noticed again until Burckhardt traveled past the temple on a journey through the desert. He noted its existence to his explorer friend Giovanni Belzoni, who made two attempts to dig out the entrance to the temple, succeeding on his second try.

One hundred and thirty years later, the Abu Simbel Temple was in danger of becoming an underwater site due to the planned construction of a dam that would have caused the waters of the Nile River to rise rapidly. Several solutions were proposed, one being the creation of an underwater viewing site for the temple. In the end, the decision was made to move the temple, block by block, to higher ground and reassemble it. The massive relocation project took four years, and the temple still stands today as a popular historical site for tourists.

Exterior sculpture detail

AJANTA CAVES

Medallion
carving on pillar

A young British military officer, stationed in India in 1819, is on a tiger-hunting expedition with some friends near the Waghora River in the Indian state of Maharashtra. The officer glances up into the dense jungle and sees what appears to be the arched entrance to a cave. He signals to his hunting party and enters the cave to find a magnificent Buddhist prayer hall filled with detailed carvings, colorful paintings, and a large statue of the Buddha. This is exactly what happened to cavalryman John Smith when he stumbled upon Cave 10, which brought the entire horseshoe-shaped chain of Ajanta Caves to the attention of the Western world for the first time.

Portion of chain of caves, exterior

INDIA

Ajanta Caves

The Ajanta Caves were cut directly into the rocky side of a mountain, and they were created in two distinct phases. Caves 9, 10, 12, 13, and 15A were built sometime between the 2nd century BC and AD 100. Beginning in the 5th century AD, construction began on Caves 1–8, 11, 14, and 15B–29; there is some debate as to the time period when the last of the caves was completed. The caves generally fall into two categories: worship halls and monasteries.

Experts say the artwork and architecture displayed in the Ajanta Caves are some of the finest surviving examples of Indian art from this time period and reflect the beautiful culture and history of India.

Interior columns and sculptures

The murals depicting the lives and rebirths
of the Buddha still glow with brilliant colors,
even centuries after they were painted.
The columns and walls separating sections
of each cave remain strong. The sculptures
both in the walls and on freestanding
statues showcase great detail and
control by the artists.

Carved
elephant

Buddha sculpture

ANGKOR

DISCOVERED 1860 DISCOVERED

Angkor Wat

Located in the Siem Reap province of Cambodia, Angkor Archaeological Park covers more than 400 square kilometers (154.4 square miles). The area served as the capital city of the Khmer civilization from approximately the 9th through the 15th centuries AD. Researchers are unsure why Angkor was abandoned as the capital.

In 1860, Henri Mouhot, a French naturalist, came upon the massive site during a trek through the Cambodian jungle. He brought Angkor to the attention of the rest of the world, and soon after, archaeologists began working in the area, seeking information about the ancient civilization. Research and restoration work continue there even in the present day; with the remains of hundreds of temples and thousands of other structures, researchers have plenty to keep them busy.

Terrace of Elephants

Bas-relief: [BAH-relief] a method of carving or sculpting where the background of a medium is carved away, leaving the figures to appear raised above the background, in three-dimensional form.

• Siem Reap
CAMBODIA

Stone faces at Angkor Thom
and Banteay Kdei Temple (left)

Sculpture at
Angkor Thom

The crown jewel of Angkor Archaeological Park is Angkor Wat (translated "temple city"). Originally built as a Hindu temple, it was being used as a Buddhist center of worship by the latter part of the 12th century. The temple is one of the premier examples of Khmer art and architecture during the time of its use.

Besides the giant temple city, the Angkor area is also home to Angkor Thom, Terrace of Elephants, Banteay Kdei, Ta Keo, and a great variety of water management, storage, and dispersion structures that were used to manage precipitation during the monsoon season. With an estimated 2.5 million guests each year, Angkor is one of the most visited tourist attractions in Cambodia and all of Asia.

The archaeological park includes miles of bas-relief carvings illustrating Indian mythology—deities, mythological animals, plants, and people—in great detail. Pediments, lintels, and walls are filled with lovely examples of Khmer artwork. Angkor Wat itself houses many intricate Buddhist statues.

DISCOVERED 1864 DISCOVERED

JERUSALEM TUNNELS

The Western Wall, also known as the Wailing Wall, is part of the longest remaining portion of the wall that once surrounded the Jewish Temple Mount in Jerusalem. King Herod built the wall around the temple in about 20 BC. The Romans later destroyed the temple in AD 70, but a portion of the retaining wall survived. Today, the Western Wall forms a part of the support structure that surrounds the Dome of the Rock and is located in the Old City of Jerusalem.

Archaeologists first began digging tunnels near the Western Wall in 1864. Among the first discoveries made through the digging of these tunnels were wells and water systems that date back to Biblical times. These archaeologists also uncovered large stone arches that once connected the city to the Temple Mount. Later, after the Six-Day War in 1967, more tunnels were dug below the wall, eventually creating the Western Wall Plaza and Tunnels, which are now popular tourist attractions. During a tour of the tunnels, tourists can touch the original stones of the temple wall and walk where the market would have been during Jesus' time, as evidenced through the thousands of ancient coins that archaeologists have found there.

ISRAEL

Jerusalem →

Tunnel passage

Broken stone cup found in excavation

Very recently, archaeologists have discovered what they believe to be a city council building from 2,000 years ago. This building, which was probably built around AD 20, could have been used as a dining room where the important people of society could stop as they made their way to the temple. Soon, this ancient city council building will be added to the Western Wall Tunnel Tours.

Lit tunnel passage

Oil lamps like this were found in the newest building, along with clay cooking vessels, a broken stone mug, and a portion of a carved pillar decoration.

Retaining wall: a wall that holds back earth or water

HISSARLIK

Ancient Greek history incorporates stories of people and places that capture our imagination. By piecing together the details of those stories, archaeologists working in Turkey discovered what is considered to be the remains of the ancient Greek city of Troy.

Historic texts described Troy's location specifically to be found along the coast of present-day Turkey where the Aegean Sea narrows into the Dardanelles. From these texts, connections were made between the geography of the area and the rubble of what was formerly known as Ilion, the inspiration for Homer's epic story *Iliad*. Excavation of the mounded site did not begin until 1865 when an amateur British archaeologist named Frank Calvert purchased a portion of the mound and began investigating. His excitement grew as he dug, but without financial backing, he decided to turn his findings over to well-known archaeologist Heinrich Schliemann of Germany.

Hissarlik TURKEY

Sanctuary walls and wells

Schliemann was better funded but less professional, and in 1870 he used dynamite to delve for treasures within the ruins of the layered mound. The discovery of a cache of gold known as "King Priam's treasure" convinced Schliemann he had indeed discovered Troy, but his assistant, William Dörpfeld, continued with a careful dig in and around the site through the 1890s. After a span of time, Carl Blegen picked up the work in the 1930s, taking scientific measures and samples of soils and artifacts throughout nine distinct layers. His findings encouraged interest with archaeological teams from two universities between 1988 and 2005 under the leadership of Manfred Korfmann.

Columns and blocks excavated from the site

Silver vessels from King Priam's treasure

Stone ramp and southwest gateway

Their work laid out a clear timeline of the transition of a small fishing village around 3,000 BC to a city of wealth and Bronze Age technology nearly 500 years later, finally developing into the legendary city of Troy inhabited by Greeks and Romans throughout the next millennium.

MESA VERDE

Spiral petroglyph

Nestled in the mountains of Colorado is a most impressive and fascinating sight: hundreds of cliff dwellings. These ancient buildings were made from wooden beams, sandstone, tiny bits of stone, and mortar made from ash, soil, and water. The Ancestral Pueblo people who inhabited these buildings were not very tall when compared to today's averages—only a little over five feet tall—and most adults lived to around just 33 years old.

These cliff dwellings, now encompassed by Mesa Verde National Park, came to the attention of the general public toward the end of the 19th century. A large amount of interest resulted in quite a bit of damage to the curious site. The largest of the dwellings was given the name of "Cliff Palace." This single dwelling has around 150 rooms! Circular excavated areas called kivas—23 of them—are also part of Cliff Palace.

In 1888, Richard Wetherill, a local to the area, began exploring the dwellings and searching for artifacts. Then Gustaf Nordenskiöld, a scholar who was visiting from Sweden, excavated around 24 dwellings, including Cliff Palace, in the summer of 1891. The artifacts that were discovered included human remains and objects used in funerals. Many of these artifacts were returned to local native tribes in 2019.

Excavation and preservation of Cliff Palace and the other dwellings continued throughout the 1900s. Today, efforts are focused on preserving this incredible historic site.

COLORADO

Mesa Verde

UNITED STATES

Petroglyph at Mesa Verde's
Petroglyph Point

MACHU PICCHU

DISCOVERED 1911 DISCOVERED

800F POSTES 2019
Machu Picchu, Pérou
REPUBLIQUE TOGOLAISE

Togolese Republic postage stamp, 2019

PERU

Machu-Picchu

In 1911, an explorer named Hiram Bingham III was searching for lost Incan cities with the help of local Peruvian farmers in the Andes Mountains. Bingham was amazed when he was led to the ruins of an empty city with over 200 buildings built into the side of a ridge between two mountains. Initially believed to be a religious training site, it was later discovered that the ancient city of Machu Picchu had been built by the Incans in the 15th century as a vacation retreat for the Incan king Pachacutec and his support staff.

Constructed from blocks of polished stone that were fitted together so precisely and tightly that no mortar was needed between them, Machu Picchu had houses, temples, storage buildings, and statues. Large, flat terraces filled with grass were created for farming, and a complex irrigation system brought water to the whole city.

Stone paths connected the different parts of the city and led away from Machu Picchu to other Incan villages, most likely to aid in the trading of supplies and goods with other cities.

Hiking path to Machu Picchu

At its peak, Machu Picchu housed around 750 people, most of them there to serve the king. The city was abandoned suddenly around the time that the Spanish came to conquer the Incas. Archaeologists theorize that either this takeover by the Spanish or a severe drought forced the city's inhabitants to leave, and the city was undisturbed and covered with jungle growth for almost 400 years until Bingham stumbled upon its remains.

Roughly 2,500 tourists per day take a train into the city of Aguas Calientes and, from there, take a bus or hike up to the ruins. Standing by the main entrance, visitors get an eagle-eye view of the breathtaking city, and then they can take a guided one-way tour through the Machu Picchu ruins.

Machu Picchu ruins

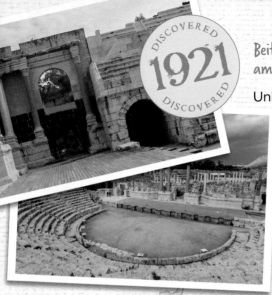

DISCOVERED 1921 DISCOVERED

Beit She'an theater and amphitheater (left)

BEIT SHE'AN

Unlike many other newly-discovered archaeological sites, the city of Beit She'an in the Galilee region of northern Israel has been continuously inhabited for 6,000 years. One of the oldest cities in present-day Israel, there is evidence that the city once housed cave dwellers and was also a busy community during the Neolithic period and the Bronze Age.

Located near the Sea of Galilee and the city of Nazareth, where Jesus grew up, Beit She'an is mentioned in the Old Testament and has been studied in depth for its significance in Biblical accounts. The city has been through many cultural transformations; it was once inhabited by ancient Canaanites, Egyptians, Israelites, and Philistines, and during the time of Jesus' life, it was an important Roman city known as Scythopolis.

Beit She'an

ISRAEL

Beginning in the 1920s and continuing through the present day, archaeologists have excavated at least twenty layers of the tel—the mound created by the accumulation of the remains of ancient civilizations. Eventually, an earthquake brought the expansive architecture to the ground, and the city never really recovered. Today, tourists can discover for themselves the history of Israel through viewing the remains from each historic period at this well-preserved ancient site.

Modern excavations have primarily focused on the artifacts and remains of the once large, thriving Roman city of Scythopolis, just south of the tel. Scythopolis had majestic pillars, a theater, an amphitheater, and other Roman luxuries, many of which have been uncovered and reconstructed.

Stone column with sculpted capital

Sculpted stone art (left);
Floor mosaic (above)

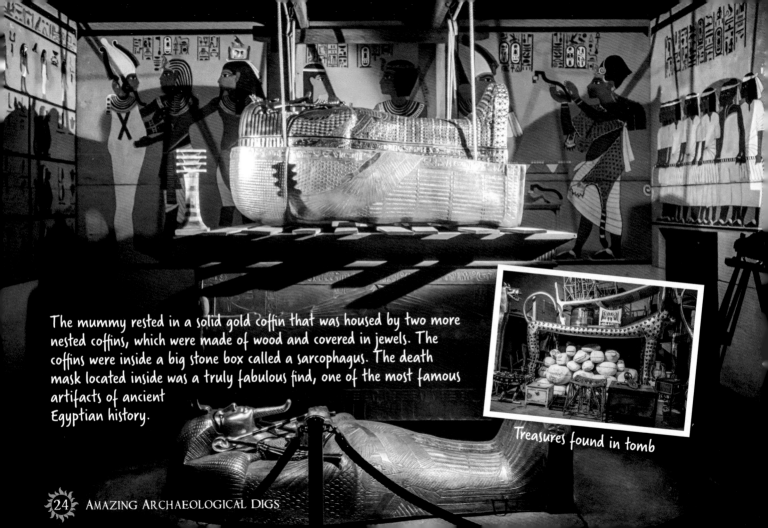

The mummy rested in a solid gold coffin that was housed by two more nested coffins, which were made of wood and covered in jewels. The coffins were inside a big stone box called a sarcophagus. The death mask located inside was a truly fabulous find, one of the most famous artifacts of ancient Egyptian history.

Treasures found in tomb

TUTANKHAMEN'S TOMB

You've probably heard about "King Tut." His tomb is one of the most famous archaeological finds to date. On November 4, 1922, Egyptologist Howard Carter discovered one step of a stairway that was buried in debris in the Valley of the Kings in Egypt. Carter and his financial backer, Lord Carnarvon, opened the tomb on November 26, 1922.

EGYPT
Valley of the Kings

Tutankhamen was a king in the 18th Egyptian dynasty. He became king when he was nine years old, and his reign was short, only from around 1332 BC to 1323 BC (less than ten years). This young king was buried with several thousand treasures and artifacts, including statues, jewelry, and clothing. It took Carter's team ten years to document the tomb's contents! Most of these items now belong to the Cairo Museum.

King Tutankhamen's death mask

The tomb was large, around 111.5 square meters (1,200 square feet) in size, and contained four rooms. In one of those rooms was the most exciting find of the entire dig: the mummified body of King Tutankhamen. The walls in the burial chamber were covered with brightly painted murals. The paintings had brown spots all over them, which were created by microbes.

Scientists aren't sure why King Tutankhamen died. He and his tomb and treasures still fascinate people today. Modern technology has even been used to construct a digital view of what the young king may have looked like!

THONIS-HERACLEION AND CANOPUS

In 1933, a pilot flew over Abu Qir Bay, an inlet of the Mediterranean Sea off the coast of Egypt. During his flight he spotted the ruins of ancient buildings under the water. Over 60 years later, archaeologists would explore these two sunken cities: Thonis-Heracleion and Canopus.

Thonis-Heracleion ("Thonis" being the Egyptian name and "Heracleion" being the Greek name) was a large city in ancient Egypt. Researchers believe it was founded during the 8th century BC. Canopus was another town located about 3.5 kilometers (2 miles) west of Thonis-Heracleion. Because these cities were so close to the sea, they suffered a combination of tsunamis, earthquakes, and rising sea levels. Eventually, the cities experienced soil liquefaction, a process that caused the soil to become liquid and the cities to sink underwater completely by the 8th century AD.

Thonis-Heracleion and Canopus

EGYPT

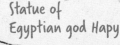
Statue of Egyptian god Hapy

Coast of Alexandria, Egypt, about 32 km (20 miles) southwest of the Thonis-Heracleion wreck site (above); statues of Arsinoe and Ptolemy II (right)

In 2000, the European Institute for Underwater Archaeology (IEASM) began exploring the underwater ruins. They found many items preserved by clay and mud, including ceramic objects, bronze ladles, coins, jewelry, and even riverboats! One incredible discovery was a statue of the Egyptian god Hapy made of pink granite; the statue, standing at 5.4 meters (17.7 feet) tall and weighing nearly 20 tons (40,000 pounds), is the tallest likeness ever found of an Egyptian god. Researchers also found pink granite statues of the pharaoh Ptolemy II and his queen Arsinoe, standing at 5 meters (16.4 feet) tall.

Because the ruins of Thonis-Heracleion and Canopus were so well preserved, even underwater, archaeologists have been able to learn much about ancient Egyptian life, especially about their religion. But there is much more work to do—Franck Goddio, the director of IEASM, estimates that only five percent of these amazing cities have been discovered so far!

SUTTON HOO

Anglo-Saxon coins

The famous Old English epic poem *Beowulf* opens with a scene in which an entire ship is buried with the body of a Danish king and his treasures inside. Sutton Hoo, in England, is a real-life version of this incredible story! This cemetery dates back to the 7th century, but excavating the 20 burial mounds only began in 1938.

Landowner Edith Pretty commissioned amateur archaeologist Basil Brown to begin digging in the burial mounds. Basil was the first to discover evidence that a ship might have been buried underground. When a professor at Cambridge University heard rumors about this major archaeological find for England, other archaeologists took over the site, and Basil Brown was told to stop excavating.

The finds included ornate weaponry, drinking cups, pieces of armor and clothing, and evidence that there was once a body, possibly that of a king, buried there. The iconic Sutton Hoo helmet was also found here.

ENGLAND

Sutton Hoo

Gold buckle (above);
Sutton Hoo shield (right)

On the brink of World War II, a new team of archaeologists uncovered the rest of the perfectly preserved form of a ship containing a trove of treasures. The wood planks had decayed, but the iron rivets remained in their original places.

Other burial mounds at Sutton Hoo were created for all classes of people, from criminals and common villagers to extremely wealthy families and royalty. There is even a grave that once held a man and his horse! The first mounds that were explored had already been vandalized and cleaned out by grave robbers.

The treasure found at Sutton Hoo is said to be one of the greatest archaeological finds in England's history. Many of the artifacts that have been found are on display at the British Museum, and tourists can see a recreation of the burial chamber and its treasures at the site of the Sutton Hoo excavation.

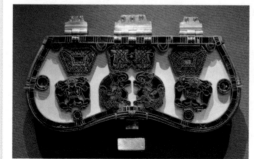

ornate purse lid

Sutton Hoo helmet

 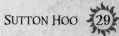

The Lascaux cave has around 6,000 paintings and engravings in the vast tunnels of the underground system. The depictions are mostly of animals—deer, horses, aurochs (an extinct type of ox), a bear, and some felines. There is also one depiction of a human, along with some mysterious geometric designs.

LASCAUX CAVE

Imagine discovering a magnificent art gallery buried underground that has not seen the light of day for thousands of years! Four teenage boys wandering around France's Dordogne region made such a discovery in 1940 when they climbed down a 15-meter (50-foot) shaft into a cave and were greeted by an array of beautiful paintings and engravings. The boys were astounded and brought their teacher back to the cave to show him their find. Lascaux cave was soon receiving thousands of visitors, which took a toll on the air quality. Algae and mold began to grow, destroying the precious art, so the cave was closed to the public in 1963. Replicas were built so the public could still safely marvel over the ancient art.

Lascaux cave is located in Southwest France near the village of Montignac. Amazingly, there are 147 other ancient sites and 25 other decorated caves in the same region! Though no one knows for sure why the paintings exist, it does not diminish the impressive parietal art.

Parietal: a term for prehistoric art found in caves, rock shelters, or cliff overhangs

FRANCE

Lascaux Cave

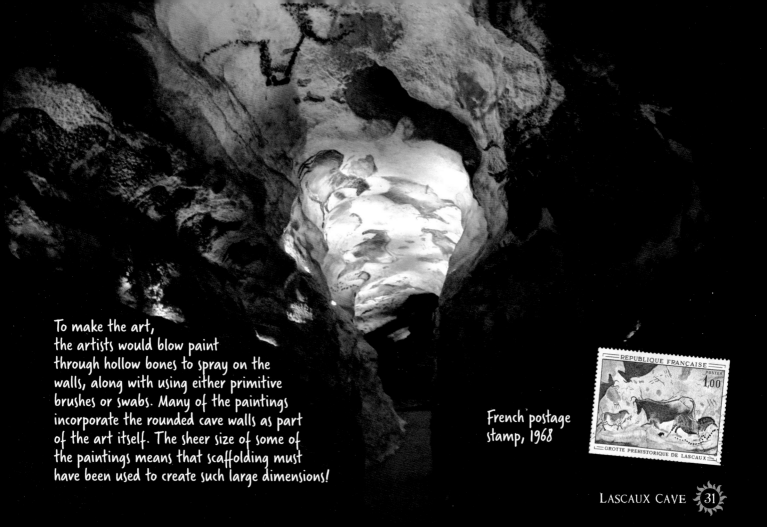

To make the art,
the artists would blow paint
through hollow bones to spray on the
walls, along with using either primitive
brushes or swabs. Many of the paintings
incorporate the rounded cave walls as part
of the art itself. The sheer size of some of
the paintings means that scaffolding must
have been used to create such large dimensions!

French postage
stamp, 1968

REPUBLIQUE FRANCAISE
POSTES
1,00
GROTTE PREHISTORIQUE DE LASCAUX

DEAD SEA SCROLLS

In 1947, a Bedouin shepherd was exploring the cliffs near the northwestern shore of the Dead Sea in Israel. He came across some caves, and inside he found pottery jars with pieces of ancient parchment. Written on the old scroll fragments were ancient texts from the time of Christ. Over the next decade, thousands of parchments were discovered. These papers collectively are what we call today the famous Dead Sea Scrolls.

The scroll fragments are old—2,000 years old. They were left behind by ancient Jews. This group of Jews, who called themselves "sons of light," worried that foreigners were corrupting Judaism. So a few withdrew to the wilderness—to this area called Qumran— to practice their religion in peace and await the Messiah.

Qumran cave where scrolls were found

What makes the Dead Sea Scrolls so significant is that they contain copies of the Bible written around the time of Jesus Christ. The scrolls contain parts of every book of the Hebrew Bible except Esther, plus some extra texts not in the Bible. Scholars can compare these texts with other ones to learn more about the history of the Bible.

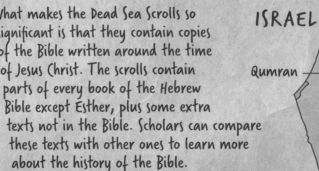

ISRAEL

Qumran

Dead Sea

Many fragments of the Dead Sea Scrolls were found in jars like these.

The scrolls also tell us about what life was like for people who lived around the time of Christ, including this unique group of Jews who lived on the shores of the Dead Sea.

One of the scrolls on display at the Shrine of the Book in Israel

Most of these scholars were men, and their society was much like that of a monastery. They lived in buildings atop the cliffs. As part of their religion, the people participated in ritual cleansings (sacred baths) and communal meals (meaning they all ate their food together). They also copied their sacred writings in a special room dedicated to writing, called a scriptorium. The finished scrolls were moved to the caves to remain safe and hidden from enemies.

Saint Aiden, founder of Lindisfarne

LINDISFARNE

Lindisfarne Priory, located on Holy Island off the coast of current-day English Northumberland, was founded by Irish monks in AD 635. It has been particularly notable several times throughout history.

About 40 years after the priory's founding, a monk named Cuthbert arrived there. He became a bishop who was well-known for his pastoral and healing gifts. St. Cuthbert was canonized after his death, and this led to the island's rise in popularity as a pilgrimage site.

Lindisfarne's prominence increased, as did its distinction as a center of Christian scholarship. The emphasis placed on learning resulted in the creation of the Lindisfarne Gospels, one of the most incredible illuminated manuscripts to survive from early medieval England. This artistic masterpiece includes the four gospels, some associated texts, and several gorgeous full-page color illustrations.

Lindisfarne

ENGLAND

Unfortunately, Lindisfarne's fame and prosperity also attracted those with less noble motives. In 793 AD, Viking pirates attacked the island, pillaging the priory and enslaving or killing many monks. It was the first significant Viking raid in that part of Europe, and it shocked the entire continent. By the 9th century, the priory was virtually abandoned. The remaining monks took St. Cuthbert's coffin and other important artifacts inland, seeking security. Today, tourists can cross over to Lindisfarne Island during low tide and tour the ruins of the castle.

Lindisfarne Priory ruins

Page from the Lindisfarne Gospels

Lindisfarne Castle

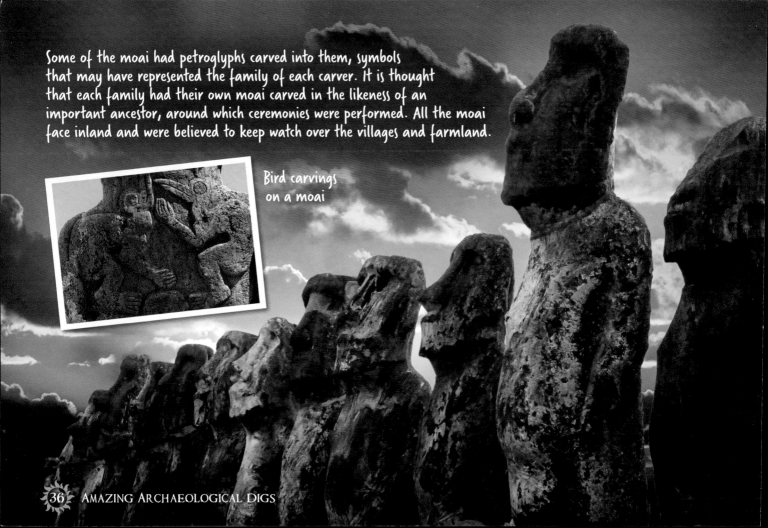

Some of the moai had petroglyphs carved into them, symbols that may have represented the family of each carver. It is thought that each family had their own moai carved in the likeness of an important ancestor, around which ceremonies were performed. All the moai face inland and were believed to keep watch over the villages and farmland.

Bird carvings on a moai

EASTER ISLAND, CHILE

In the South Pacific Ocean, about 2,300 miles off the coast of Chile, lies the fascinating and mysterious Easter Island, called Rapa Nui by its native peoples. This 163-square-kilometer (63-square-mile) volcanic island is famous for hundreds of giant figures called moai [MOH–i]. You might be familiar with the unique heads protruding from the ground, but there is much more below the surface. Over the years from 1955 to the present, archaeologists carefully dug around the moai and began to unearth the hidden bodies on which the heads sit.

Although the massive statues were carved into lightweight porous rock formed by volcanic ash, the largest moai weighs over 80 tons! We don't know for sure how they were moved onto their sacred platforms, called ahu, but popular theories include that they were slid down the hills or rocked back and forth using ropes tied around their heads.

Easter Island moai

Other archaeological findings include hundreds of multi-purpose tools and wonderful art within a network of caves formed by crashing waves and volcanic activity. The cave paintings often depict marine life, faces, and most commonly a mythical Birdman. Most include a red pigment that was also used in rituals around the moai.

Recently, protective coatings were added to many of the moai to shield the soft rock from weathering and erosion, and the cliffs on which many stand have been reinforced. Hopefully, these efforts will preserve Rapa Nui's history for many years to come.

Warrior detail

TERRACOTTA WARRIORS

On March 29, 1974, about 20 miles from the city of Xi'an in north-central China, a farmer began digging a new well. As he shoveled the dirt aside, he made an amazing discovery. There, under the dirt, he found life-sized statues of soldiers made of clay. Archaeologists realized that he had found the burial site of the first emperor of China, Qin Shi Huang.

When Emperor Qin ruled China over 2,000 years ago, he believed he would need a palace to live in and an army to serve him after death. Around 720,000 workers spent almost 40 years building not only the palace and the pyramid mound under which the emperor was buried, but also offices, storehouses, stables, and an army of warriors made of clay.

CHINA   Xi'an

8,000 warriors guard the site.

There are around 8,000 unique, life-sized warriors guarding the burial site. Each one was made from terracotta, a reddish-brown clay. Their heads, arms, and bodies were molded separately and then assembled. Next, wet clay was smoothed over the surface, and artists modeled each face to give them individual features and expressions. Their different hair, clothes, and gestures give clues about each soldier's rank, job, and social status. Each statue was baked for a long time at high heat in a large, outside oven (called a kiln) until it was hard and durable. Finally, they were each painted with bright colors and given real bronze weapons.

Around 246 BC Emperor Qin ordered his workers to build an army to protect him after death. Now, archaeologists study these ancient statues and the other parts of his burial site to learn more about life in China during his reign.

Terracotta horses stand with the warriors.

Each warrior has unique features and expressions.

BRITAIN'S POMPEII: MUST FARM

First discovered in 1999, Must Farm, or "Britain's Pompeii," gives archaeologists an exciting look into Britain's Bronze Age. In the scope of history, the Bronze Age refers to the time period when humans started to use metal, specifically bronze (made by combining copper and tin), to make tools and weapons. The Bronze Age in Britain began around 2500 BC and continued until 800 BC, when people started using iron more often.

Located near a quarry at Whittlesey, England, Must Farm is an especially great find for archaeologists because the homes and items there are very well-preserved. As Pompeii was destroyed by fire from an erupting volcano, Must Farm was burned down, possibly by invading warriors. However, the flames of the fire were soon extinguished as the homes there quickly collapsed and fell into the river below.

In 2016, archaeologists at Must Farm uncovered a wheel that is 3,000 years old! This fascinating find is the oldest and most complete wheel that has been found in Britain. The discovery of this wheel speaks to the sophistication of the people at Must Farm, shedding more light on the Bronze Age and its technology.

Must Farm wheel

ENGLAND

Must Farm

The houses were round structures built on stilts, and near their remains, archaeologists have discovered many everyday items. Among them are wooden utensils; wooden boxes; platters; pottery, such as bowls and cups; metalwork, such as axes and razors; weapons; and textiles, all preserved incredibly well after being covered with layers of silt in the riverbed.

Pottery excavated from the site

The textiles have been particularly interesting to archaeologists. The fabrics found here are of a fine quality, some made with thin threads, similar in diameter to human hair. Another interesting find is glass beads, which are believed to have come from the Middle East or the Mediterranean.

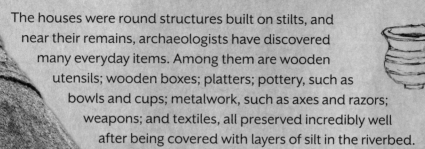

Beads made of amber and jet lignite have been found at Must Farm.

Amber

Jet lignite

Replica iron age house, similar to the houses scientists believe existed at Must Farm

ROOSWIJK

A Dutch East Indiaman ship, similar in appearance to the Rooswijk, drawn by Francis Swaine, circa 1770

2005

Early in the year 1740, the Dutch East India Company sent a ship loaded with goods for sale or trade on its second voyage to the island of Jakarta in the East Indies. The ship, called the *Rooswijk*, set sail from Amsterdam in the Netherlands for its 6- to 8-month journey. Only one day into the voyage, a fierce storm forced the ship aground in the Goodwin Sands off the southeastern coast of England.

It is uncertain how many passengers were on board, but all passengers, as well as the 237 crew members, lost their lives that day. In 2005, the wreck was discovered by an amateur diver, and some of the silver was salvaged and turned over to the Dutch government for analysis. There was interest in the site, but before physically exploring the wreckage, archaeologists and scientists from both Historic England and the Dutch government used various technological scans to prioritize which items and areas to investigate.

ENGLAND

Goodwin Sands

In 2017 and 2018, items were lifted from the seabed to be identified and cataloged. What has excited historians about this underwater dig is that the contents of the ship remain mostly intact, and this is the first such ship to be scientifically excavated. The ship was heavy and sank quickly, and the shifting sandy seafloor calcified, contained, and preserved the majority of the sunken items.

Among the items discovered were Mexican silver, sheet metal, iron nails, glass beads and bottles, stone columns, and mail to be delivered to Jakarta.

Chevron-patterned glass beads and glass onion bottles (similar to the ones shown here) were recovered. Thousands of silver coins also were found, including Mexican-minted Spanish Reales, known as pieces of eight.

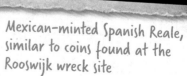

Mexican-minted Spanish Reale, similar to coins found at the Rooswijk wreck site

In the field of archaeology, the discovery of our past as a people continues. History is constantly being rewritten with the unearthing of new artifacts, buildings, artwork, remains of once-living beings, tools, and written work. It is the way we interpret these findings that is important, and the way we use that information to shape human behavior in the future that is crucial to making the future as bright as possible.

As archaeologist Sarah Parcak, who specializes in Egyptology, once said:

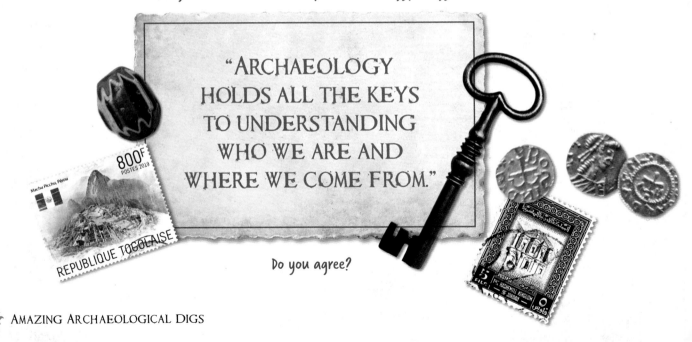

"ARCHAEOLOGY HOLDS ALL THE KEYS TO UNDERSTANDING WHO WE ARE AND WHERE WE COME FROM."

Do you agree?